100 MIND-BLOWING FACTS ABOUT SPACE

Exploring the Wonders of the Cosmos

Welcome to "100 Mind-Blowing Facts about Space"! In the vast expanse of the universe, there exist countless wonders and mysteries waiting to be explored. This book is a journey through the cosmos, presenting you with a collection of awe-inspiring and mind-boggling facts about space.

Before we embark on this celestial adventure, it is essential to acknowledge that the universe is an ever-evolving entity, and our understanding of it is constantly expanding. The knowledge we possess today is the result of years of scientific exploration, observation, and analysis. However, we must remember that science is a dynamic and progressive field, continuously uncovering new discoveries and revising our existing understanding.

The facts presented in this book are based on the current state of scientific knowledge up until 2023. They have been rigorously researched and verified by reputable sources within the scientific community. However, it is important to recognize that as scientific advancements continue, some of the information presented here may evolve or be refined in the future.

The beauty of science lies in its capacity to adapt and refine its theories as new evidence emerges. What we know today may be supplemented, modified, or even challenged by future findings. Embracing this dynamic nature of scientific understanding allows us to remain open-minded and receptive to new ideas.

"100 Mind-Blowing Facts about Space" serves as a snapshot of our current understanding of the cosmos, a testament to the remarkable knowledge we have amassed thus far. As we delve into each fact, we invite you to marvel at the wonders of space and ponder the endless possibilities that lie beyond our planet.

Remember, the cosmos is a vast and ever-changing tapestry, and it is through the exploration of space that we continue to push the boundaries of human knowledge. So, embark on this celestial journey with an open mind, ready to be amazed, and let the wonders of the universe unfold before you.

Fact 1: The Moon is slowly drifting away from Earth.

Our beloved Moon, Earth's only natural satellite, is gradually moving away from us. As it orbits around our planet, tidal forces caused by the Moon's gravity create a transfer of energy, gradually pushing it to a higher orbit. This means that in the distant future, generations to come might witness a night sky without the familiar presence of the Moon.

Fact 2: Venus spins in the opposite direction to most other planets.

While most planets in our solar system spin counterclockwise on their axes, Venus chooses to go against the flow. This means that if you were to stand on Venus, the Sun would rise in the west and set in the east. The reason behind this peculiar phenomenon remains a mystery, captivating scientists and astronomers alike.

Fact 3: A day on Mercury is longer than its year.

Surprisingly, a single day on the planet Mercury, the closest planet to the Sun, lasts longer than its entire year. It takes about 176 Earth days for Mercury to complete a full rotation on its axis, while it takes approximately 88 Earth days for it to orbit the Sun. Imagine experiencing sunrises and sunsets at such a leisurely pace!

Fact 4: Jupiter has a massive storm called the Great Red Spot.

Jupiter, the largest planet in our solar system, boasts a colossal storm known as the Great Red Spot. This storm has been raging for centuries, and it is so enormous that three Earths could fit within its boundaries. Its swirling, crimson clouds and turbulent nature make it a truly mesmerizing feature of our cosmic neighborhood.

Fact 5: A day on Saturn lasts only about 10.7 hours.

If you're a fan of short days, Saturn would be your ideal celestial destination. This gas giant, famous for its majestic rings, completes a full rotation on its axis in just 10.7 hours. So, while we may spend 24 hours in a day on Earth, Saturn's inhabitants would have considerably less time to tackle their daily activities.

Fact 6: Neptune's blue color comes from methane gas.

Neptune, the icy giant located farthest from the Sun in our solar system, is famous for its striking blue hue. This captivating color is a result of its atmosphere containing traces of methane gas. Methane absorbs red light, allowing the blue light to reflect back and grace Neptune with its alluring cerulean shade.

Fact 7: The Sun is so large that a million Earths could fit inside it.

Our radiant Sun, the star at the center of our solar system, is truly massive. Its vast size is so incredible that you could fit approximately one million Earths inside it! Its tremendous gravitational pull keeps our planetary family in its captivating orbit, nurturing life on our beloved planet.

Fact 8: There are more stars in the universe than grains of sand on all the Earth's beaches.

Imagine standing on a beach, surrounded by countless grains of sand beneath your feet. Now, try to fathom that there are more stars in the universe than all those tiny sand particles combined. This mind-boggling fact reminds us of the sheer enormity and wonder of our ever-expanding cosmos.

Fact 9: Mars has the largest volcano in the solar system.

Mars, the red planet, boasts an incredible volcano called Olympus Mons. Standing at an astonishing height of around 13.6 miles (22 kilometers), it is the tallest volcano in the entire solar system. Its colossal size and the potential mysteries hidden within its ancient volcanic past make it a truly captivating feature of our neighboring planet.

Fact 10: Astronauts on the International Space Station witness 16 sunrises and sunsets in a day.

Life aboard the International Space Station (ISS) is nothing short of extraordinary. Orbiting the Earth at high speed, astronauts on the ISS witness an astonishing 16 sunrises and sunsets every day. Imagine the awe-inspiring views and the sense of wonder that must accompany each new dawn and twilight on their celestial home away from home.

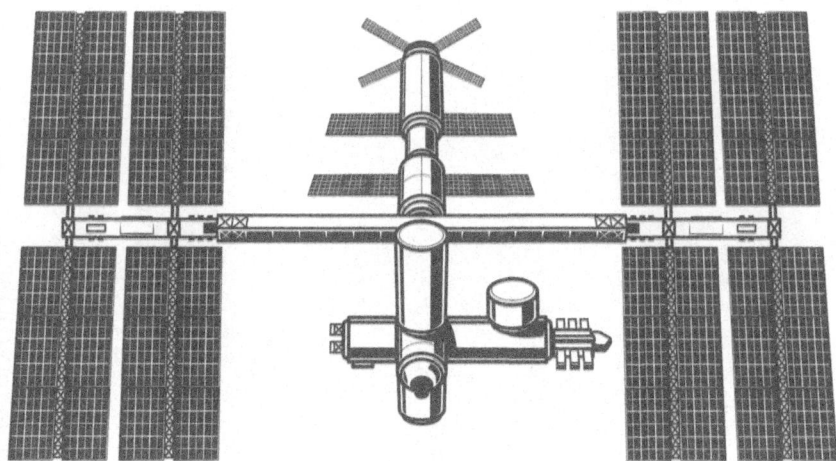

Fact 11: The hottest planet in our solar system is not the closest to the Sun.

Although it may seem counterintuitive, the hottest planet in our solar system is not the one closest to the Sun. Instead, that title goes to Venus, despite it being the second planet from the Sun. Venus' thick atmosphere traps heat like a greenhouse, creating a scorching environment with temperatures reaching a blistering 900 degrees Fahrenheit (475 degrees Celsius), hotter than Mercury.

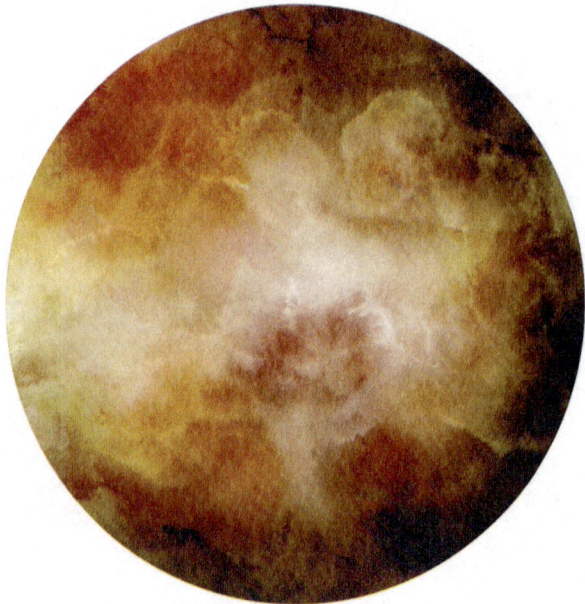

Fact 12: Saturn's rings are made up of countless ice particles.

Saturn's magnificent rings, one of the most iconic features of the planet, are composed of billions upon billions of ice particles. These particles range in size from tiny grains to enormous chunks, spanning a mind-boggling distance of more than 175,000 miles (280,000 kilometers). The enchanting beauty and intricate structure of Saturn's rings continue to captivate scientists and stargazers alike.

Fact 13: The largest black hole discovered is billions of times more massive than our Sun.

Black holes, enigmatic cosmic entities with immense gravitational pull, come in various sizes. The largest black hole known to science is located in the galaxy Messier 87 and is estimated to have a mass billions of times greater than that of our Sun. The mind-blowing power and mysterious nature of black holes make them a subject of fascination and ongoing scientific exploration.

Fact 14: A teaspoonful of a neutron star would weigh about a billion tons on Earth.

Neutron stars are the remnants of massive stars that have undergone a supernova explosion. These incredibly dense celestial objects pack a mind-boggling amount of mass into a relatively small volume. In fact, a mere teaspoonful of neutron star material would weigh roughly a billion tons on Earth, highlighting the extraordinary gravitational forces at play.

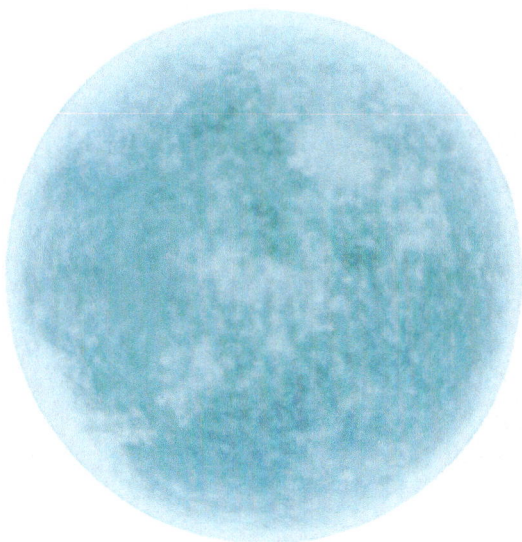

Fact 15: The first exoplanet was discovered in 1992.

Until relatively recently, the existence of planets outside our solar system was purely speculative. However, in 1992, astronomers made a groundbreaking discovery by identifying the first exoplanet, a planet orbiting a star other than our Sun. This momentous finding opened the floodgates to a new era of exoplanet exploration, revolutionizing our understanding of planetary systems beyond our own.

Fact 16: The center of the Milky Way harbors a supermassive black hole.

Nestled in the heart of our galaxy, the Milky Way, lies a supermassive black hole known as Sagittarius A*. With a mass equivalent to about four million Suns, it exerts an immense gravitational pull on surrounding matter. This enigmatic cosmic behemoth remains shrouded in mystery, leaving astronomers eager to unlock its secrets.

Fact 17: Astronauts height can increase while in space.

Spending time in the microgravity environment of space can lead to a peculiar phenomenon known as "space-induced growth." Without the constant downward pull of gravity compressing their spines, astronauts can experience a slight increase in height while in space. However, this height gain is temporary and vanishes upon their return to Earth's gravity.

Fact 18: The majority of the universe is made up of mysterious dark matter and dark energy.

Ordinary matter, the stuff we interact with every day, accounts for only a small fraction of the universe. The majority of the cosmos is believed to be composed of mysterious entities called dark matter and dark energy. While their exact nature remains elusive, scientists hypothesize that dark matter provides the gravitational scaffolding for the formation of galaxies, while dark energy accelerates the expansion of the universe itself.

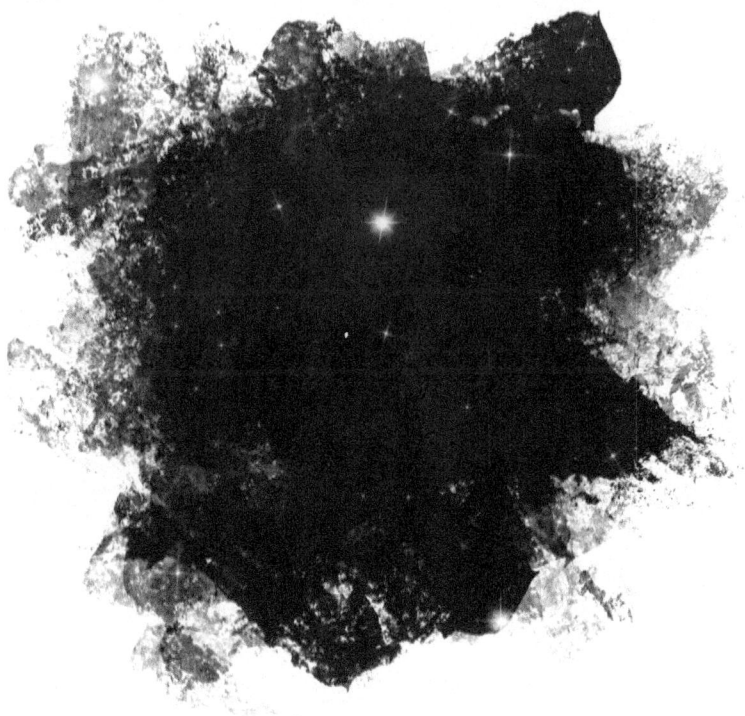

Fact 19: The first woman to travel to space was Valentina Tereshkova.

On June 16, 1963, Valentina Tereshkova, a Soviet cosmonaut, became the first woman to venture into space. Her groundbreaking mission aboard the Vostok 6 spacecraft paved the way for female astronauts to follow in her footsteps. Tereshkova's extraordinary achievement shattered gender barriers, inspiring generations of women to pursue careers in space exploration.

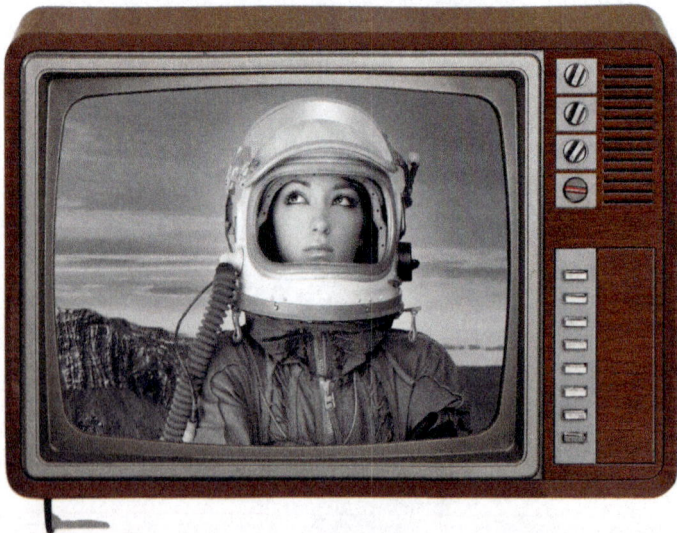

Fact 20: The Andromeda galaxy is hurtling towards our Milky Way.

Brace yourself for a cosmic collision! Our neighboring galaxy, Andromeda, is not just a distant object in the night sky. It is hurtling toward our Milky Way at a speed of about 110 kilometers per second (68 miles per second). However, don't worry too much—it's still billions of years away, and the eventual merger will likely result in a stunning dance of cosmic proportions.

Fact 21: Uranus rotates on its side, giving it unique seasons.

Uranus, the seventh planet from the Sun, possesses a peculiar characteristic—it spins on its side. This extreme axial tilt of about 98 degrees gives Uranus its distinct seasons. Imagine experiencing years where the Sun never sets or rises, followed by periods of constant sunlight. The intriguing nature of Uranus's rotational axis continues to pique the curiosity of astronomers.

Fact 22: The Voyager 1 spacecraft has left our solar system.

Launched in 1977, the Voyager 1 spacecraft embarked on a grand journey through the depths of space. In 2012, it became the first human-made object to enter interstellar space, venturing beyond the boundaries of our solar system. Voyager 1 carries with it a golden record, containing sounds and images representing Earth's diverse cultures and life forms, an homage to our existence in the vastness of the cosmos.

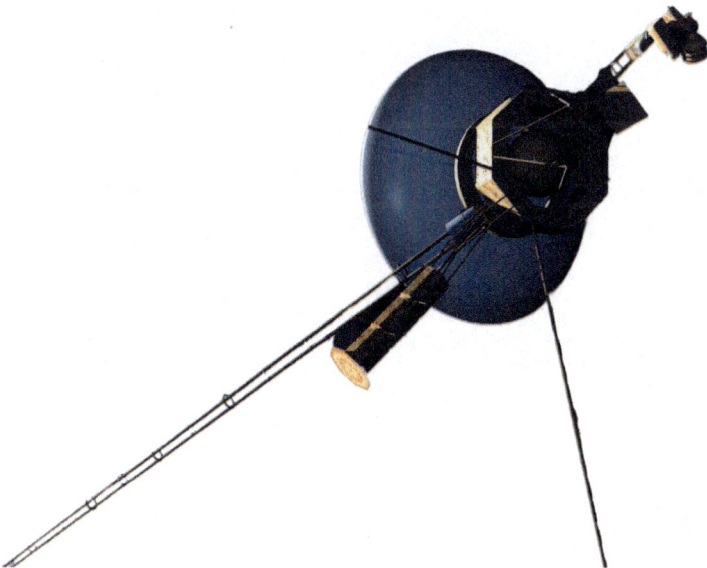

Fact 23: The Sun's energy comes from nuclear fusion.

The Sun, our celestial powerhouse, derives its energy through a process called nuclear fusion. In the Sun's core, hydrogen atoms fuse together under immense heat and pressure, forming helium and releasing an enormous amount of energy in the process. This energy sustains life on Earth and illuminates our days, reminding us of the profound forces at work in the heart of our star.

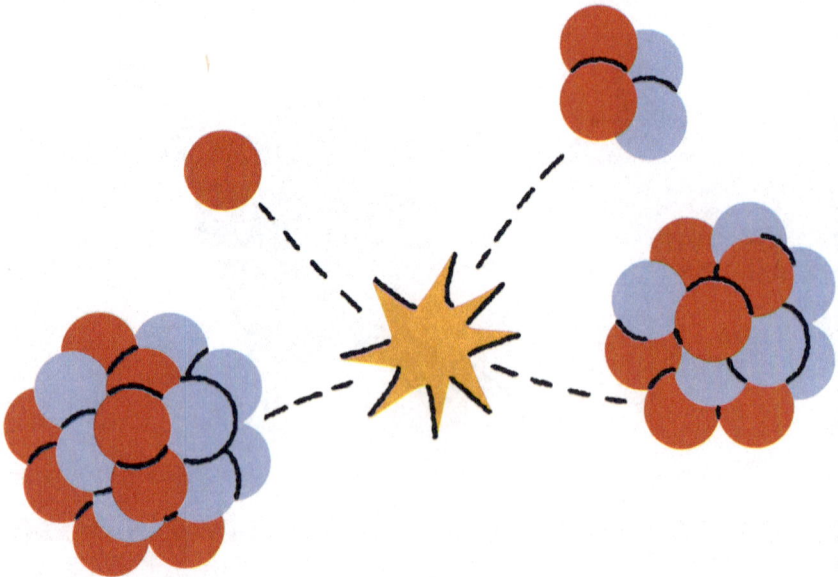

Fact 24: The Crab Nebula is the remnants of a supernova observed in 1054 AD.

The Crab Nebula, located in the constellation Taurus, is a stunning celestial object born from a colossal explosion. In the year 1054 AD, astronomers from various cultures observed a sudden burst of light—the result of a supernova detonation. The remnants of this cataclysmic event continue to captivate stargazers today, reminding us of the ever-changing nature of our universe.

Fact 25: The Kuiper Belt is home to icy objects beyond Neptune's orbit.

Beyond the orbit of Neptune lies a vast region of icy objects known as the Kuiper Belt. This region serves as a celestial archive, preserving remnants from the early stages of our solar system's formation. Among the inhabitants of the Kuiper Belt is Pluto, once considered the ninth planet before being reclassified as a dwarf planet, sparking a reevaluation of our definitions of planetary status.

Fact 26: Black holes can distort time and space itself.

Black holes are not just objects with immense gravitational pull—they also bend and warp the fabric of spacetime itself. Their gravitational influence is so profound that they create a gravitational well, where time slows down and space becomes severely distorted. The mind-bending effects of black holes continue to challenge our understanding of the fundamental laws of physics.

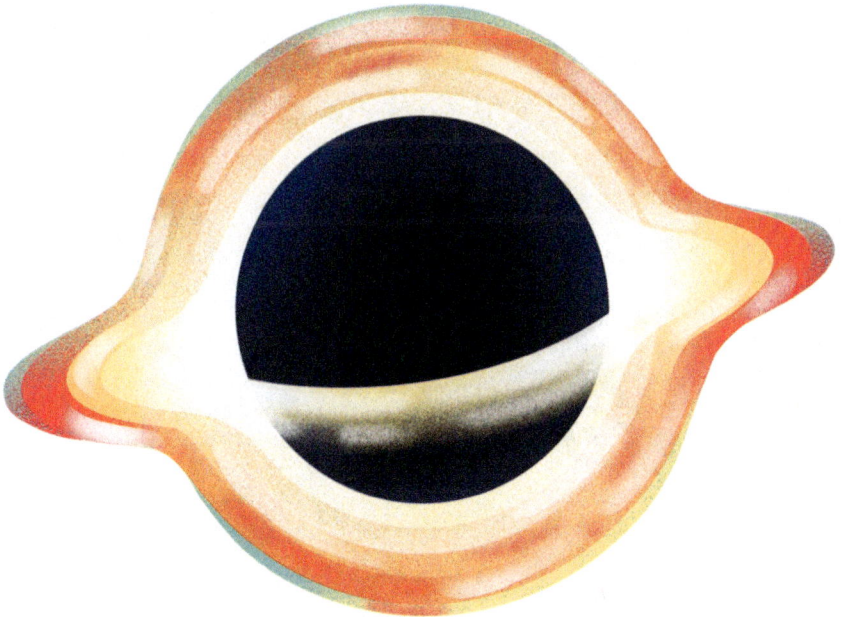

Fact 27: The Hubble Space Telescope has captured stunning images of deep space.

The Hubble Space Telescope, launched in 1990, has revolutionized our understanding of the cosmos. From its vantage point in space, it has captured breathtaking images of distant galaxies, colorful nebulae, and awe-inspiring cosmic phenomena. Its images have not only expanded our scientific knowledge but also ignited a sense of wonder and awe in people around the world.

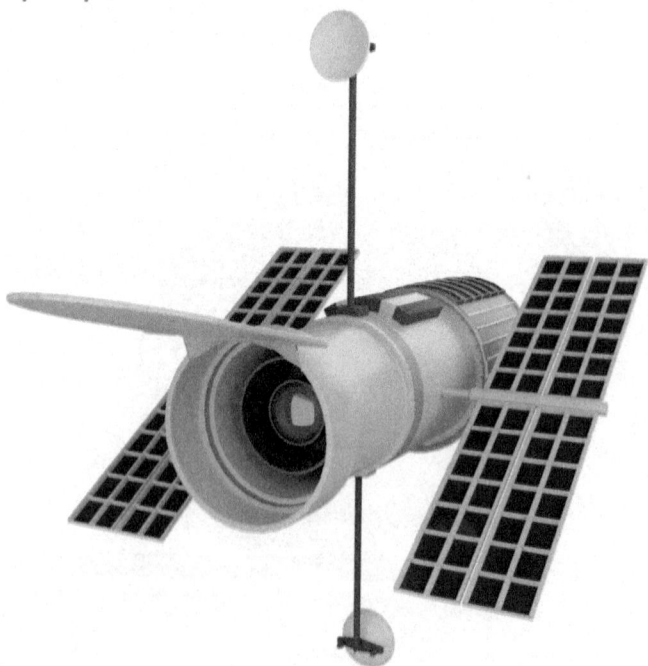

Fact 28: Astronauts experience a lack of taste and smell in space.

Life aboard a space mission is filled with incredible experiences, but it also comes with its quirks. Astronauts often report a decrease in their sense of taste and smell while in space. The microgravity environment affects their nasal passages, making it difficult for aromas to reach their olfactory receptors. As a result, even the most delicious-sounding space meals may not have the same flavorful punch they do on Earth.

Fact 29: Mars has the deepest canyon in the solar system.

Mars is a planet of superlatives. In addition to Olympus Mons, the tallest volcano, it also boasts Valles Marineris, the deepest canyon in the entire solar system. This massive chasm stretches over 2,500 miles (4,000 kilometers) in length and delves to depths of up to 6.8 miles (11 kilometers). Exploring these colossal features would be a dream come true for any adventurous space explorer.

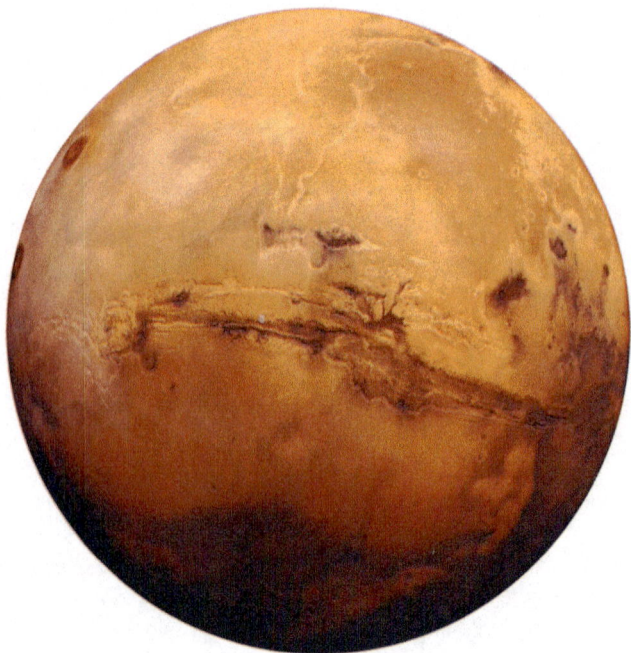

Fact 30: Space is not completely empty, it contains a low density of particles.

While space is often imagined as an empty void, it is not devoid of matter entirely. Although incredibly sparse, space contains a low density of particles, including atoms, molecules, and even dust grains. This interstellar medium, though tenuous, plays a crucial role in the formation of stars and the cosmic processes that shape our universe.

Fact 31: The universe is expanding, and the farther away a galaxy is, the faster it moves away from us.

The concept of an expanding universe was one of the most significant discoveries in cosmology. Observations have revealed that galaxies are receding from us at speeds proportional to their distance. This suggests that space itself is stretching, carrying distant galaxies away from us at an accelerating pace. The ongoing expansion of the universe is a testament to its dynamic and ever-changing nature.

Fact 32: Astronauts can experience bone and muscle loss in microgravity.

Life in microgravity poses unique challenges to the human body. Without the constant tug of gravity, astronauts' bones and muscles experience a decline in strength and density. To counteract this, astronauts engage in rigorous exercise routines while in space, ensuring that their bodies remain strong and resilient despite the weightless environment.

Fact 33: The concept of 'time dilation'
means time can pass differently for
observers in different gravitational fields.

The theory of relativity introduced the concept of "time dilation," revealing that time can flow at different rates depending on the strength of the gravitational field experienced by an observer. This means that clocks in stronger gravitational fields tick slower compared to those in weaker fields. The mind-bending nature of time dilation underscores the intricate relationship between gravity, spacetime, and the passage of time.

Fact 34: The Oort Cloud is a hypothetical region of icy objects beyond the Kuiper Belt.

The Oort Cloud is a theoretical sphere-shaped region surrounding our solar system, located far beyond the Kuiper Belt. It is believed to be a vast reservoir of comets and icy objects, extending several light-years from the Sun. Although its existence is inferred from observations, direct evidence of the Oort Cloud remains elusive. It represents a potentially rich area of exploration for future space missions.

Fact 35: Stars come in a wide range of sizes, from tiny brown dwarfs to colossal supergiants.

Stars, those luminous beacons that dot our night sky, come in a remarkable range of sizes. At one end of the spectrum, we have brown dwarfs, often referred to as "failed stars," which are too small to sustain nuclear fusion in their cores. At the other end, we have supergiants, colossal stars many times larger than our Sun. The incredible diversity of star sizes adds to the tapestry of cosmic wonders awaiting our exploration.

Fact 36: The James Webb Space Telescope is the largest, most powerful telescope ever launched into space.

The James Webb Space Telescope (JWST), was launched in 2021, and revolutionized our understanding of the universe. With a segmented primary mirror spanning 6.5 meters (21.3 feet) in diameter, it will be the largest telescope ever launched into space. The JWST's advanced instruments and cutting-edge technology are enabling astronomers to observe distant galaxies, planetary systems, and more with unprecedented detail and clarity.

Fact 37: The International Space Station (ISS) is the largest human-made structure in space.

Orbiting about 250 miles (400 kilometers) above the Earth, the International Space Station (ISS) is an extraordinary engineering feat. It is the largest human-made structure in space, spanning the size of a football field and consisting of multiple modules that house living quarters, laboratories, and docking ports for visiting spacecraft. The ISS serves as a vital research platform and a symbol of international collaboration in space exploration.

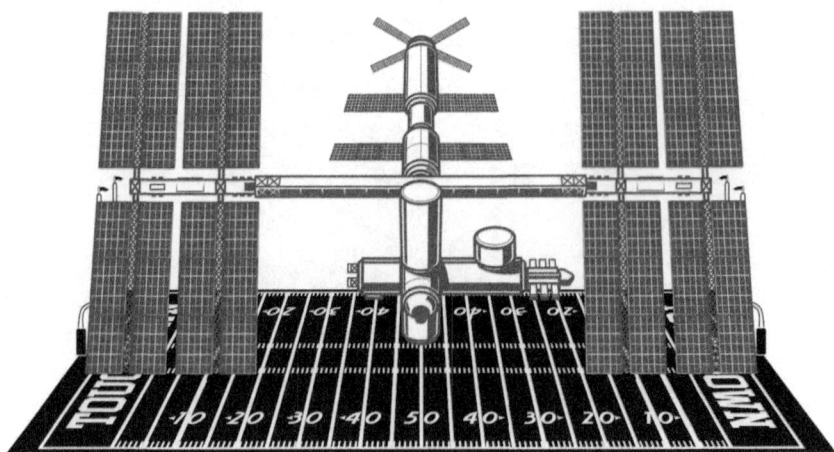

Fact 38: The Crab Pulsar emits regular pulses of radiation hundreds of times per second.

The Crab Pulsar, located at the heart of the Crab Nebula, is a highly magnetized neutron star that emits regular pulses of radiation at incredibly fast rates. It spins on its axis hundreds of times per second, making it one of the fastest-spinning celestial objects known to astronomers. The Crab Pulsar's rapid pulses and the complex dynamics of its surrounding nebula provide a captivating window into the fascinating world of pulsars.

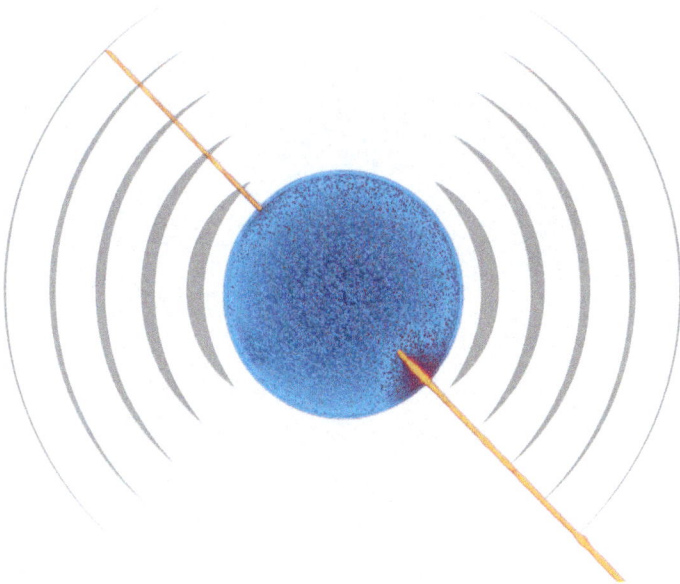

Fact 39: Astronomers use gravitational lensing to study distant galaxies.

Gravitational lensing is a phenomenon where the gravitational pull of a massive object, such as a galaxy or a black hole, bends the path of light from more distant objects. This distortion acts as a cosmic magnifying glass, allowing astronomers to observe and study distant galaxies that would otherwise be too faint or distant to detect. Gravitational lensing has provided us with breathtaking insights into the nature of the universe and its distant inhabitants.

Fact 40: The concept of 'stellar nucleosynthesis' explains how stars forge elements in their cores.

Stars are cosmic alchemists, responsible for the creation of elements that make up the universe. Stellar nucleosynthesis is the process by which stars fuse lighter elements into heavier ones in their cores, releasing energy in the process. This remarkable phenomenon explains how elements like carbon, oxygen, and iron, crucial for the formation of planets and life as we know it, are forged in the fiery hearts of stars.

Fact 41: Earth magnetic field acts as a shield against the solar wind.

Earth is protected by a magnetic field, a powerful invisible force that extends into space and shields us from the solar wind, a stream of charged particles emitted by the Sun. The interaction between Earth's magnetic field and the solar wind gives rise to stunning auroras, while also preserving our atmosphere and protecting life on our planet from the potentially harmful effects of solar radiation.

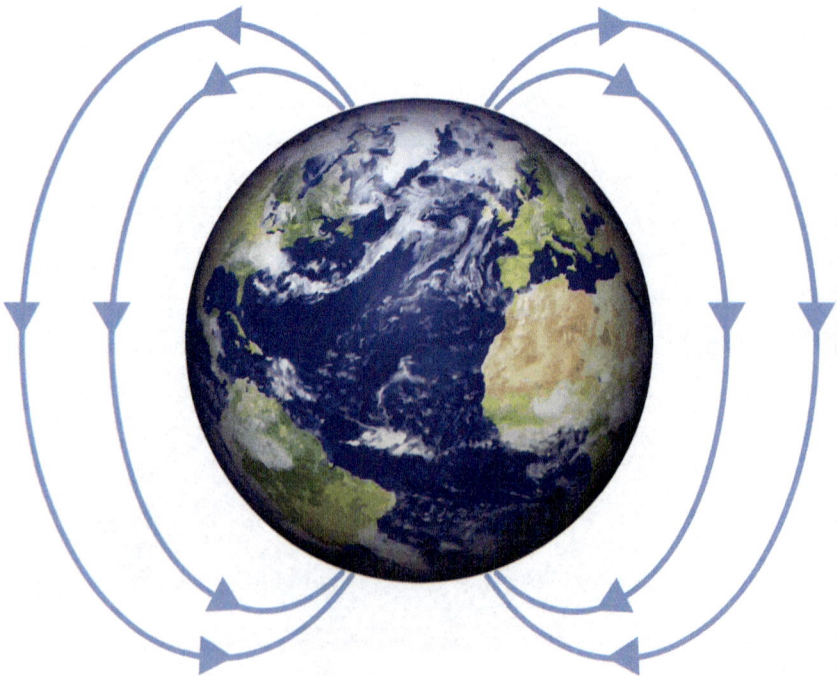

Fact 42: The concept of 'gravitational waves' confirms the existence of ripples in the fabric of spacetime.

Gravitational waves are ripples in the fabric of spacetime itself, caused by the acceleration of massive objects. These waves were predicted by Albert Einstein's general theory of relativity and were detected for the first time in 2015. The discovery opened up a new window for observing the universe, as gravitational waves carry unique information about cataclysmic cosmic events, such as the merger of black holes and neutron stars.

Fact 43: The phenomenon known as 'stellar cannibalism' occurs when a star devours its neighboring star.

In the cosmic dance of celestial bodies, stars sometimes engage in a not-so-friendly encounter. Stellar cannibalism, also known as stellar or tidal disruption, occurs when a star gravitationally rips apart and devours a neighboring star. This dramatic event releases a burst of energy and results in the formation of an accretion disk, where matter spirals towards the surviving star. Stellar cannibalism showcases the intense gravitational forces at play in the universe.

Fact 44: The concept of 'supernovae' refers to the explosive deaths of massive stars.

Supernovas are among the most cataclysmic events in the universe, marking the explosive deaths of massive stars. When these stellar giants exhaust their nuclear fuel, they undergo a runaway fusion process or collapse, resulting in a spectacular explosion. Supernovae release vast amounts of energy and create heavy elements, dispersing them into space. These stellar fireworks play a vital role in the life cycles of galaxies and the formation of new stars and planetary systems.

Fact 45: The concept of 'cosmic microwave background radiation' is a remnant of the early universe

The cosmic microwave background radiation (CMB) is a faint, uniform radiation that permeates the entire universe. It is a relic from the early stages of the universe, just 380,000 years after the Big Bang. The CMB offers a snapshot of the universe's infancy, providing crucial insights into its composition, evolution, and the seeds that led to the formation of galaxies and cosmic structures we see today.

BIG BANG

Fact 46: Astronauts experience a phenomenon called 'space adaptation syndrome' or 'space sickness' in microgravity.

The transition from Earth's gravity to the weightless environment of space can have a profound impact on astronauts' bodies. Many astronauts experience space adaptation syndrome, commonly known as space sickness, which can cause symptoms such as nausea, disorientation, and headaches. The effects of microgravity on the human vestibular system, which helps maintain balance, contribute to this temporary adjustment period. Understanding space sickness is crucial for ensuring the well-being of astronauts during extended space missions.

Fact 47: Black holes can emit powerful jets of particles and radiation from their vicinity.

While black holes are known for their ability to engulf everything, they can also launch powerful jets of particles and radiation into space. These jets, often associated with supermassive black holes at the centers of galaxies, can extend over vast distances and release incredible amounts of energy. The precise mechanisms behind the formation and launching of these jets are still subjects of active research, highlighting the enigmatic nature of black holes.

Fact 48: Saturn's moon, Titan, has lakes and seas of liquid methane and ethane on its surface.

Titan, one of Saturn's moons, is unique in our solar system. It has lakes and seas on its surface, but instead of water, they are filled with liquid methane and ethane. Titan's frigid temperatures and hydrocarbon-rich environment create a unique landscape that scientists find intriguing and reminiscent of Earth's hydrological cycle.

Fact 49: The red color of Mars is due to iron oxide, also known as rust, covering its surface.

Mars is often referred to as the "Red Planet" because of its distinctive reddish hue. This coloration is caused by iron oxide, commonly known as rust, which covers the planet's surface. The iron-rich minerals on Mars oxidize over time, giving the Martian soil and rocks their characteristic reddish appearance.

Fact 50: Jupiter has a powerful magnetic field that is 14 times stronger than Earth's.

Jupiter, the largest planet in our solar system, boasts a powerful magnetic field. In fact, Jupiter's magnetic field is approximately 14 times stronger than Earth's. This immense magnetic field creates intense radiation belts around the planet, and its interaction with Jupiter's moons generates fascinating phenomena, such as the spectacular auroras on Jupiter.

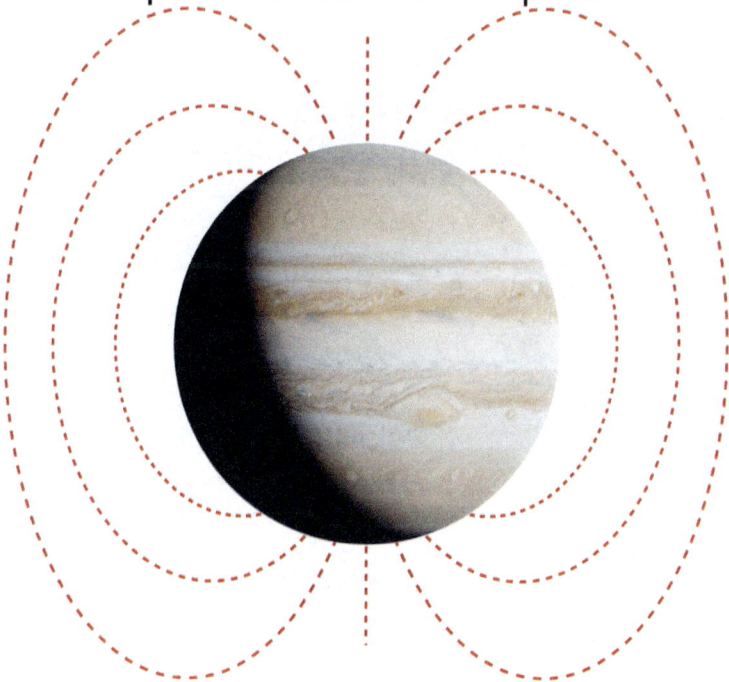

Fact 51: Neptune experiences the strongest winds in the solar system.

Neptune, the eighth planet from the Sun, is known for its fierce winds. The planet experiences the strongest winds in our solar system, with speeds reaching up to 1,500 miles per hour (2,400 kilometers per hour). These high-velocity winds create dynamic weather patterns and contribute to the striking blue appearance of Neptune.

Fact 52: The surface gravity on a white dwarf is so strong that a teaspoon of its material would weigh as much as an elephant.

White dwarfs are the remnants of low- to medium-mass stars that have exhausted their nuclear fuel. Despite their small size, their surface gravity is incredibly strong. In fact, the gravity is so intense that a teaspoon of material from a white dwarf would weigh as much as an elephant on Earth. This highlights the incredible density and gravitational forces present in these stellar remnants.

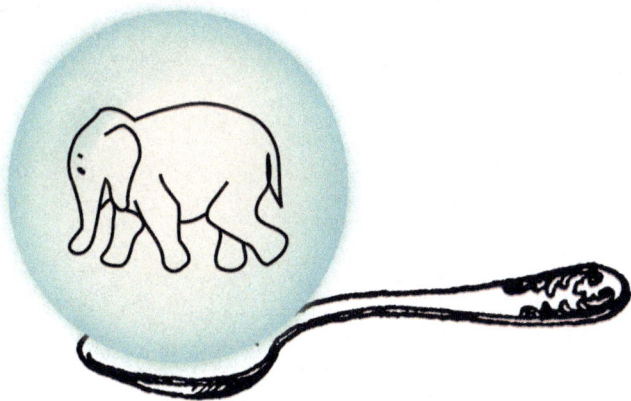

Fact 53: The Hubble Space Telescope can capture images with such clarity that it could distinguish two fireflies at a distance of 7,000 miles (11,000 kilometers)

The Hubble Space Telescope has revolutionized our understanding of the universe with its high-resolution images. It can capture light with exceptional clarity, enabling it to distinguish fine details. In fact, its imaging capabilities are so precise that it could discern two fireflies at a distance of 7,000 miles (11,000 kilometers), showcasing the incredible precision of this remarkable space observatory.

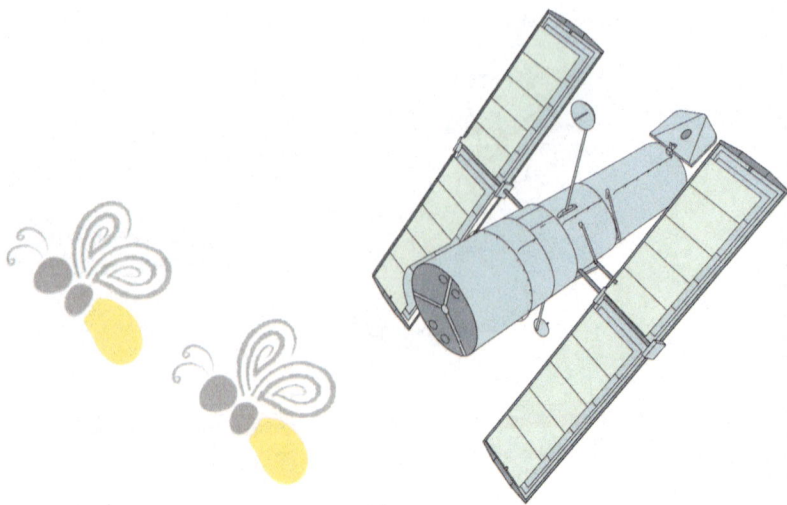

Fact 54: A year on Pluto is equivalent to about 248 Earth years.

Pluto, has an incredibly long orbital period. It takes approximately 248 Earth years for Pluto to complete one orbit around the Sun. This means that a year on Pluto is significantly longer than a human lifetime, highlighting the vast timescales at play in our solar system.

Fact 55: The largest celestial body in the asteroid belt, Ceres, is so big that it is considered a dwarf planet.

Ceres is the largest object in the asteroid belt located between Mars and Jupiter. It is so sizable that it meets the criteria to be classified as a dwarf planet. Ceres's size and composition make it an intriguing object for study, as it may harbor valuable insights into the early solar system and the formation of planets.

Fact 56: Comets are icy bodies that originate from the outer regions of the solar system.

Comets are celestial objects composed of a mixture of ice, dust, and rocky material. They originate from the outer regions of the solar system and follow elongated orbits that can bring them closer to the Sun. As a comet approaches the Sun, the heat causes the ice to vaporize, creating a glowing coma (a hazy envelope around the nucleus) and often developing distinct tails that can extend for millions of miles.

Fact 57: The concept of 'quasars' refers to extremely luminous objects powered by supermassive black holes at the centers of galaxies.

Quasars, short for "quasi-stellar objects," are incredibly bright and distant cosmic phenomena. They are powered by the accretion of mass onto supermassive black holes at the centers of galaxies. The intense gravitational forces generate enormous amounts of energy, making quasars some of the most luminous objects in the universe. Exploring quasars allows scientists to investigate the growth and evolution of galaxies.

Fact 58: The Pioneer 10 and 11 spacecraft carry a message plaque with information about Earth, intended for any extraterrestrial intelligence that may encounter them.

The Pioneer 10 and 11 spacecraft, launched in the early 1970s, were the first human-made objects to venture into the outer regions of our solar system. As a potential means of communication with extraterrestrial civilizations, both spacecraft carry a message plaque. The plaque contains information about Earth, including a map showing our location within the Milky Way galaxy, in hopes of reaching intelligent beings elsewhere in the cosmos.

Fact 59: The Apollo moon missions brought back a lot of moon rocks.

The Apollo missions, conducted by NASA in the late 1960s and early 1970s, resulted in the first and only human footsteps on the moon. The astronauts collected a remarkable 842 pounds (382 kilograms) of moon rocks and soil samples. These samples have since been studied extensively, providing valuable insights into the moon's geological history, its composition, and the processes that shaped its surface over billions of years.

Fact 60: The Martian day, also known as a sol, is only about 39 minutes and 35 seconds longer than a day on Earth.

Mars, often referred to as the "Red Planet," has a rotation period slightly longer than Earth's. A Martian day, called a sol, lasts approximately 24 hours, 39 minutes, and 35 seconds. This slight difference in day length between Earth and Mars makes it intriguing to consider the potential challenges and adjustments that would be necessary for human colonization or extended stays on the Martian surface.

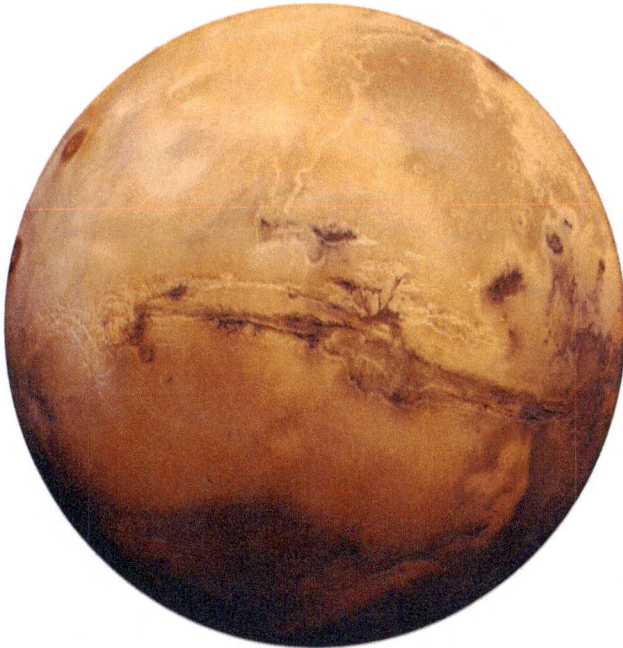

Fact 61: Saturn, known for its stunning rings, has more than 80 moons, with Titan being its largest and most intriguing satellite.

Saturn, the second-largest planet in our solar system, boasts a remarkable ring system that has fascinated astronomers for centuries. Additionally, Saturn is home to an extensive family of moons. With over 80 known moons, each with its own characteristics, Saturn's moons provide a diverse array of objects for scientific study. Among these, Titan stands out as the largest moon and possesses a dense atmosphere, making it an enticing target for future exploration.

Fact 62: The "Great Wall of Galaxies" is a massive cosmic structure consisting of multiple galaxy superclusters stretching across hundreds of millions of light-years.

The Great Wall of Galaxies is an awe-inspiring cosmic structure that spans vast distances in the universe. It is formed by a collection of galaxy superclusters, which are clusters of galaxies gravitationally bound together on an even larger scale. The Great Wall of Galaxies stretches across immense cosmic distances, spanning hundreds of millions of light-years. Its existence highlights the intricate and interconnected nature of the cosmic web, where galaxies and galaxy clusters are organized into immense filaments, sheets, and walls.

Fact 63: Europa, one of Jupiter's moons, is believed to have a subsurface ocean of liquid water beneath its icy crust.

Scientists have gathered compelling evidence suggesting that Europa harbors a global subsurface ocean of liquid water. This intriguing moon's icy crust is thought to conceal an ocean that contains more than twice the amount of water found on Earth. The existence of liquid water, combined with the potential for chemical energy and heat sources, makes Europa one of the most promising candidates for extraterrestrial life within our solar system.

Fact 64: There are enormous clouds of water vapor, spanning thousands of light-years, drifting in interstellar space.

In the vast expanse of interstellar space, astronomers have made a remarkable discovery: colossal clouds of water vapor spanning thousands of light-years. These immense clouds contain staggering amounts of water, surpassing the quantity present in Earth's oceans billions of times over. The presence of such vast reservoirs of water highlights the abundance of this essential compound throughout the cosmos and raises intriguing questions about the potential for water-based life forms beyond our planet.

Fact 65: Plumes of water vapor erupting from Europa's surface was discovered by the Hubble Space Telescope.

Observations made by the Hubble Space Telescope revealed the presence of water vapor plumes erupting from Europa's surface. These plumes provide strong evidence that the subsurface ocean may periodically vent into space. The discovery of these plumes offers a unique opportunity for future missions to directly sample and analyze the composition of Europa's subsurface ocean without the need to land on the moon's surface. It raises exciting possibilities for studying the potential habitability and signs of life within this hidden ocean.

Fact 66: There are Rogue planets, that wander through space.

Rogue planets are intriguing celestial wanderers that exist independent of any star. These solitary objects are not orbiting any star and instead roam through space freely. They are believed to have either been ejected from their original star systems or formed through gravitational collapse without ever becoming gravitationally bound to a star. Rogue planets travel through the darkness of space, devoid of the warmth and illumination provided by a parent star.

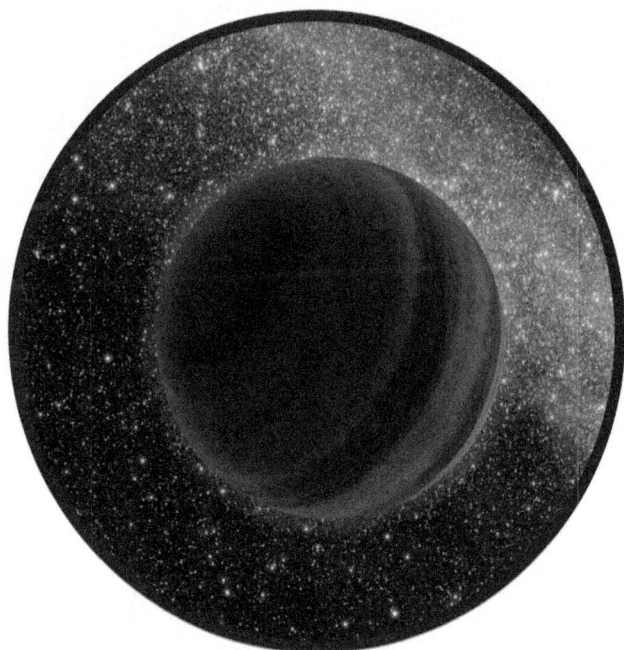

Fact 67: When we look up at the night sky, we are actually observing a picture from the past.

The light emitted by stars and celestial objects travels at a finite speed, known as the speed of light. This means that the light we see from objects in the night sky has taken a certain amount of time to reach us. For example, if a star is located 100 light-years away, the light we see from that star today actually left the star 100 years ago. In this way, the night sky serves as a beautiful and awe-inspiring cosmic time capsule, allowing us to witness the past and gain glimpses into the history of the universe.

Fact 68: The number of trees on Earth far exceeds the number of stars in the observable universe.

While it may seem astonishing, the estimated number of trees on Earth is significantly greater than the number of stars in the observable universe. The Earth is home to billions of trees across diverse ecosystems, ranging from dense forests to individual trees scattered across vast landscapes. In contrast, the observable universe contains an estimated 100 billion to 200 billion galaxies, each consisting of billions or even trillions of stars. This fact underscores the remarkable abundance and importance of trees on our planet, highlighting their essential role in sustaining life and maintaining Earth's ecological balance.

Fact 69: On Neptune, it rains diamonds.

While we commonly associate rain with water on Earth, the atmosphere of Neptune presents a unique phenomenon. Scientists believe that deep within Neptune's atmosphere, intense pressure and extreme temperatures cause carbon atoms to compress and crystallize, forming diamond particles. These diamond raindrops, made of carbon, fall towards the planet's core due to the gravitational pull. The conditions on Neptune, with its gaseous atmosphere composed mostly of hydrogen and helium, provide a remarkable environment where diamonds are formed and precipitate as "rain." This fascinating occurrence demonstrates the extraordinary diversity of conditions found in our solar system.

Fact 70: It takes approximately 8 minutes and 20 seconds for light from the Sun to travel to Earth.

Light, the fastest known phenomenon in the universe, travels at a speed of about 299,792 kilometers per second (or approximately 186,282 miles per second). Despite this incredible speed, it still takes a finite amount of time for light to travel across vast distances. In the case of the Sun, located about 149.6 million kilometers (or 93 million miles) away from Earth on average, it takes approximately 8 minutes and 20 seconds for light to reach our planet.

Fact 71: You can fit between Earth and the Moon every planet in the solar system with room to spare.

The vastness of space can be mind-boggling, but when it comes to the size of our solar system, the distances can sometimes surprise us. Despite the immense scale of the planets, their combined sizes are still smaller than the distance between the Earth and the Moon. This means that if you were to place all the planets, including Mercury, Venus, Mars, Jupiter, Saturn, Uranus, and Neptune, in a line, they would fit snugly within the space separating us from our nearest celestial neighbor, the Moon.

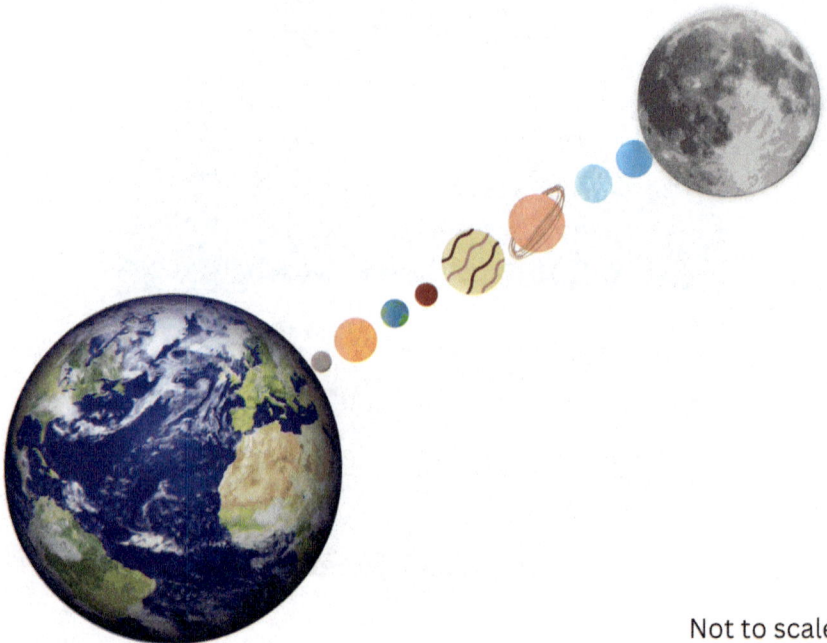

Not to scale

74

Fact 72: Io, one of Jupiter's moons, is the most volcanically active body in our solar system, with over 400 active volcanoes.

Io's volcanic activity sets it apart as the most volcanically active body in our solar system. This moon experiences intense tidal forces generated by Jupiter's gravitational pull and the gravitational interactions with other Galilean moons, causing its interior to flex and heat up. These tidal forces drive volcanic activity, with over 400 active volcanoes erupting on Io's surface. The volcanic eruptions on Io spew out a variety of materials, including sulfur, sulfur dioxide, and molten silicate lava, creating a dynamic and ever-changing landscape. The study of Io's volcanism provides valuable insights into the geologic processes and extreme environments found on this remarkable moon.

Fact 73: The sunset on Mars appears blue.

Unlike Earth, where sunsets are often characterized by warm tones of red, orange, and pink, the sunset on Mars showcases a striking blue hue. This unique phenomenon occurs due to the composition of the Martian atmosphere. The thin atmosphere of Mars contains dust particles that scatter sunlight in a different way compared to Earth's atmosphere. As the sunlight passes through the Martian atmosphere, it undergoes a scattering process known as Rayleigh scattering, which causes shorter blue wavelengths to scatter more than longer red wavelengths.

Fact 74: If you can spot the Andromeda Galaxy with your naked eyes, you can see something 14.7 billion billion miles away.

The Andromeda Galaxy, also known as M31, is a neighboring spiral galaxy to our Milky Way. It is located approximately 2.537 million light-years away from Earth. When you look up at the night sky and spot the Andromeda Galaxy without the aid of telescopes or binoculars, you are witnessing a remarkable feat. The light that reaches your eyes has traveled a staggering distance of about 14.7 billion billion miles (or 23.7 billion billion kilometers).

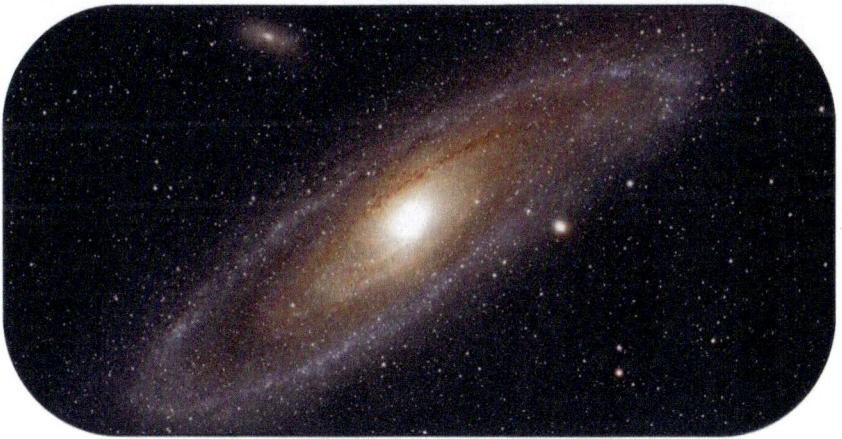

Fact 75: "Apparent size equality".

Despite the vast difference in size between the Sun and the Moon, they appear almost identical in size when viewed from Earth. This is due to the precise alignment of their apparent sizes in our sky. The Sun's diameter is about 400 times larger than that of the Moon, but it is also positioned roughly 400 times farther away. As a result, the Sun and the Moon create a stunning visual effect during a total solar eclipse, where the Moon perfectly covers the Sun, leading to the appearance of a brilliant solar corona surrounding the lunar silhouette. This celestial coincidence of apparent size equality between the Sun and the Moon is a captivating phenomenon that has fascinated and inspired humans throughout history.

Fact 76: Space is completely silent.

Unlike on Earth, where sound waves can travel through air or water and reach our ears, space is essentially a vacuum devoid of air and matter. Sound waves are mechanical vibrations that require a medium to propagate. In the absence of a medium, such as air or water, there is no material for sound to travel through in space. As a result, space is eerily silent, and astronauts in space missions rely on other means of communication, such as radio waves, to transmit and receive information.

Fact 77: A full NASA space suit is very expensive.

NASA's space suits, known as Extravehicular Mobility Units (EMUs), are sophisticated and highly specialized garments designed to protect astronauts during spacewalks and extravehicular activities (EVAs). These suits are meticulously engineered to provide life support systems, thermal regulation, communication capabilities, and protection against the harsh conditions of space. The cost of a full NASA space suit, including the development, testing, and maintenance, is estimated to be around $12,000,000.

Fact 78: The footprints left by astronauts on the Moon's surface will endure for an astonishing 100 million years.

The Moon's environment is vastly different from Earth's. On Earth, wind, rain, and other geological processes constantly reshape the surface, erasing evidence of human activity over time. However, the Moon lacks a significant atmosphere and weathering processes, which means the footprints left by astronauts during the Apollo missions remain remarkably preserved. Without the forces of erosion, these footprints are expected to persist for an estimated 100 million years, serving as enduring reminders of humanity's incredible journey to the lunar surface.

Fact 79: If two pieces of the same type of metal touch in space, they have the potential to permanently bond together.

In the vacuum of space, where there is no atmosphere or moisture to interfere, certain metals have the remarkable ability to bond when they come into contact with each other. This phenomenon is known as cold welding or contact welding. On Earth, metals naturally develop thin oxide layers on their surfaces that prevent direct bonding. However, in the absence of an atmosphere in space, these oxide layers are absent, allowing the atoms at the interface of the metals to diffuse and form atomic bonds. This can result in a strong and permanent connection between the two pieces of metal. Cold welding has been observed during space missions and is a crucial consideration when designing spacecraft and equipment to prevent unintended bonding.

Fact 80: The gas giant Jupiter is a failed star.

Jupiter, the largest planet in our solar system, is often called a failed star due to its similarities with stars in terms of its composition and structure. Like stars, Jupiter is primarily composed of hydrogen and helium, the same elements that make up stars. However, Jupiter didn't gather enough mass during its formation to reach the critical threshold necessary for nuclear fusion to occur in its core. Nuclear fusion is the process that powers stars by converting hydrogen into helium and releasing tremendous amounts of energy. Without sustained fusion, Jupiter remains a gas giant planet rather than a fully ignited star.

Fact 81: Enceladus, one of Saturn's moons, is the most reflective body in the solar system.

Enceladus possesses a remarkably high reflectivity, making it the most reflective body known in our solar system. The moon's icy surface is composed of a layer of water ice mixed with other materials. This icy surface acts like a mirror, reflecting nearly all of the sunlight that reaches it, which makes it a very cold world with temperatures around -201 °C (-330 °F). The high reflectivity of Enceladus contributes to its dazzling appearance and makes it a captivating object for scientific study and exploration.

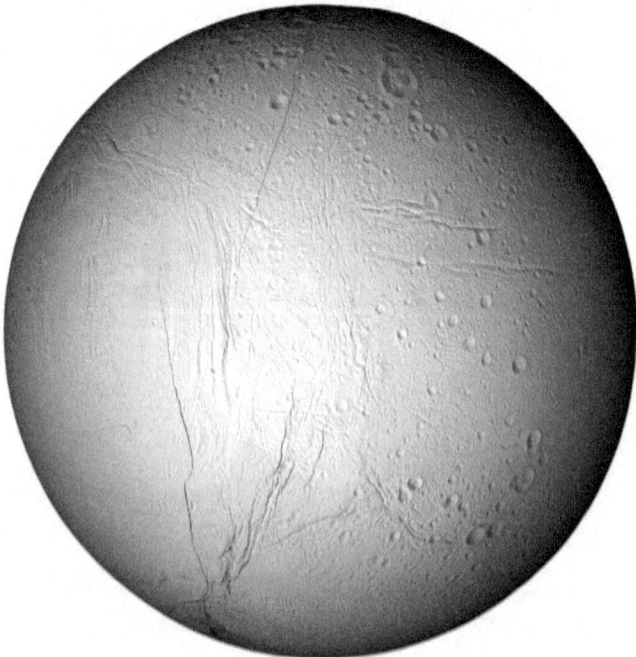

Fact 82: The Sun will engulf Earth 5 billion years from now.

As the Sun exhausts its hydrogen fuel in its core, it will enter the final stages of its life. In this phase, known as the red giant phase, the Sun will expand in size, becoming hundreds of times larger than its current size. This expansion will cause the Sun to engulf the inner planets, including our home planet Earth.

Fact 83: There are approximately 2,000,000,000,000 galaxies in the observable universe.

The vastness of the universe is truly awe-inspiring, and the number of galaxies it contains is staggering. Astronomers estimate that there are around 2 trillion galaxies in the observable universe. Each galaxy, like our own Milky Way, can contain billions to trillions of stars, along with various other celestial objects. Considering the immense number of galaxies, it emphasizes the mind-boggling scale of the cosmos and the countless mysteries waiting to be explored.

Fact 84: All known planets in the solar system have been visited by uncrewed spacecraft.

Throughout the history of space exploration, humanity has successfully sent uncrewed spacecraft to explore and study all known planets in our solar system. From Mercury to Neptune, each planet has been the target of various missions designed to investigate their atmospheres, surfaces, and unique characteristics. These robotic missions have provided us with invaluable data, stunning images, and a deeper understanding of the diverse worlds that make up our cosmic neighborhood. The achievements of these uncrewed spacecraft have paved the way for future human exploration and have expanded our knowledge of the planets in our solar system.

Fact 85: Voyager 1 and 2 have been operating for more than 40 years.

Voyager 1 and Voyager 2 are iconic spacecraft that were launched to explore the outer reaches of our solar system. Despite their launch over four decades ago, both spacecraft are still operational and transmitting valuable data back to Earth. Voyager 1 became the first human-made object to reach interstellar space in 2012, while Voyager 2 is currently on its way to interstellar space. These resilient spacecraft have provided groundbreaking insights into the outer planets, their moons, and the vastness of space, enriching our understanding of the universe.

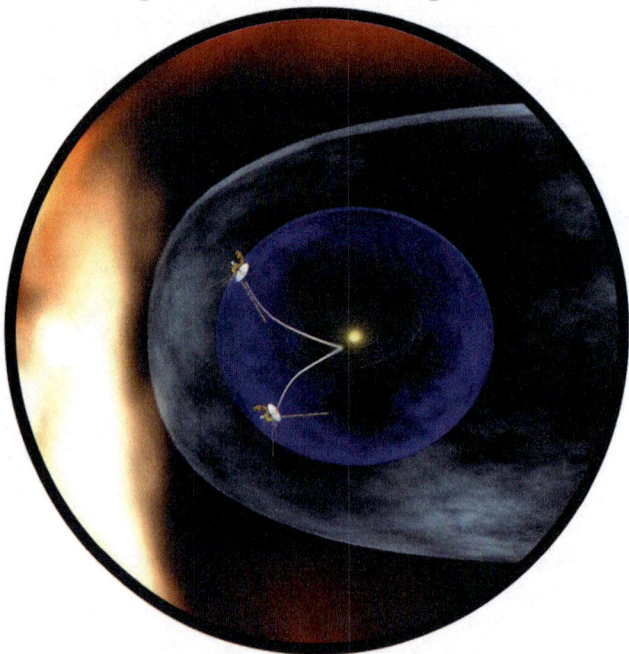

Fact 86: The closest star system to us, Proxima Centauri, is located approximately 4.25 light-years away from our solar system.

Proxima Centauri holds the distinction of being the nearest star system to our own. Situated about 4.25 light-years away, it is part of the Alpha Centauri system, which also includes the binary star system of Alpha Centauri A and B. Proxima Centauri is a small, low-mass red dwarf star that is slightly smaller and cooler than our Sun. Its proximity makes it an intriguing target for future exploration and potential investigations of exoplanets that may orbit this neighboring star.

Alpha Centauri

Fact 87: There are more than 4,000 known exoplanets

Thanks to advancements in observational techniques and space missions dedicated to exoplanet exploration, astronomers have identified and confirmed the existence of over 4,000 exoplanets. These planets orbit stars other than our Sun and vary in size, composition, and distance from their host stars. The ongoing search for exoplanets using methods like the transit and radial velocity techniques has opened up a vast and diverse exoplanetary landscape, revolutionizing our understanding of planetary systems and the potential for habitable worlds beyond our own.

Fact 88: Other planets, and even a moon, have auroras too.

Auroras, also known as the Northern and Southern Lights on Earth, are not exclusive to our planet. Other celestial bodies in our solar system, such as Jupiter, Saturn, and even a moon like Ganymede, also exhibit spectacular auroras. Just like on Earth, these auroras occur when charged particles from the Sun interact with a planet or moon's magnetic field and atmosphere. The result is a mesmerizing display of colorful lights dancing across the sky.

Fact 89: Shooting stars are space debris that burn up when they enter Earth's atmosphere.

Shooting stars, also known as meteors, are celestial objects that originate from space and enter Earth's atmosphere. Most shooting stars are small fragments of rock or dust called meteoroids, which are remnants from comets or asteroids. As a meteoroid travels through the atmosphere, the intense heat generated by its rapid entry causes it to burn up and vaporize, creating a glowing trail of light known as a meteor or shooting star. The friction between the meteoroid and the air molecules leads to its incandescent display, captivating viewers on the ground.

Fact 90: The outer space is very cold.

In the vast expanse of outer space, temperatures plummet to extremely frigid levels, with temperatures often reaching near absolute zero, which is around -273 degrees Celsius (-459 degrees Fahrenheit). Unlike Earth, which benefits from the warmth of the Sun's rays, space is a vacuum, devoid of air and thermal conduction. As a result, heat cannot be transferred efficiently, leading to an absence of warmth. The average temperature in space hovers just above absolute zero, where atoms and molecules come to a near standstill. This bone-chilling coldness is one of the many extreme conditions that astronauts and space probes must contend with during space exploration.

Fact 91: A human could survive Up To 30 seconds In Space.

In the vacuum of space, there is no air or atmospheric pressure to sustain human life. Because of that, if your lungs were full of air, they would burst. As long as there is no air in your lungs, you could survive a maximum of 30 seconds in space with no protective equipment. Within seconds, the absence of oxygen and atmospheric pressure would cause the fluids in the body, such as saliva and tears, to vaporize. The extreme cold, extreme heat, and radiation would pose immediate threats to the body. Despite the ability to survive for a short period of time due to the body's own internal pressure and reserves, the lack of oxygen and the intense environment would ultimately prove fatal.

Fact 92: There Are Laws In Space.

Space is not a lawless void. Over time, a framework of laws and treaties has been developed to regulate activities in space and promote cooperation among nations. The United Nations Office for Outer Space Affairs (UNOOSA) plays a significant role in coordinating international efforts and fostering collaboration. Key international agreements include the Outer Space Treaty, which prohibits the placement of weapons of mass destruction in space and establishes principles for peaceful exploration, and the Rescue Agreement, which outlines procedures for the rescue of astronauts in distress.

Fact 93: Mercury Is Shrinking.

Mercury, is the smallest planet in our solar system. Recent observations and data from NASA's MESSENGER spacecraft have revealed that Mercury is experiencing a phenomenon called "global contraction." Over time, the planet's interior cools and contracts, causing its surface to deform and creating prominent geological features known as scarps. These scarps are long, steep cliffs that extend for hundreds of kilometers. The shrinkage of Mercury's surface provides valuable insights into the dynamic processes occurring within the planet and sheds light on its geological evolution throughout history.

Fact 94: The Sun Is not Yellow.

Although the Sun often appears yellow when viewed from Earth, it is not inherently yellow in color. The true color of the Sun is white. The reason it appears yellow to our eyes is due to the Earth's atmosphere scattering shorter blue and green wavelengths of light more than longer red and yellow wavelengths. As a result, the sunlight reaching our eyes appears to be predominantly yellow. However, when observed from space or through specialized solar filters, the Sun appears as a white or slightly bluish-white star.

Fact 95: Stars Are Multicolored.

Stars exhibit a wide range of colors, from blue and white to yellow, orange, and red. These colors provide important clues about various aspects of a star's nature. The temperature of a star determines its color, with hotter stars appearing bluish-white and cooler stars exhibiting a reddish hue. The color of a star can also indicate its age and evolutionary stage. Additionally, the presence of specific elements in a star's atmosphere influences its color, allowing astronomers to study the chemical composition of stars by analyzing their colors.

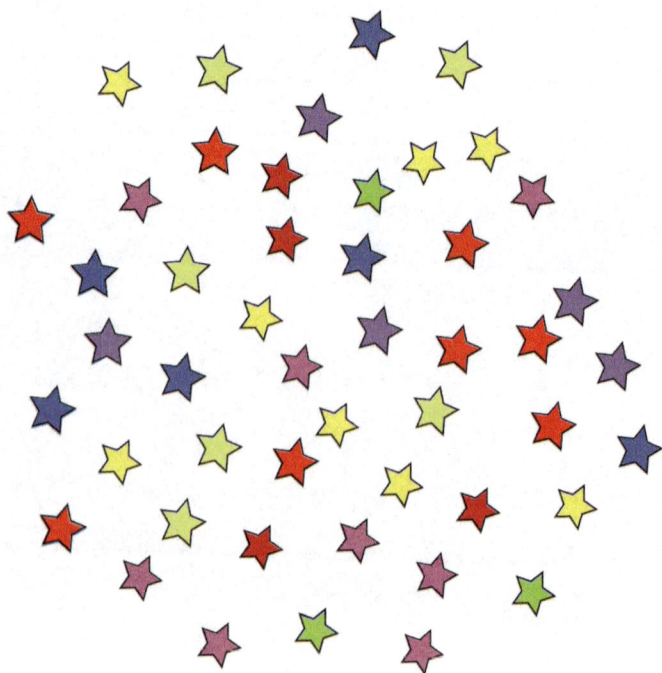

Fact 96: There Is No Dark Side of the Moon.

The idea of a dark side of the Moon often stems from the misconception that one hemisphere is permanently shrouded in darkness. However, the Moon's rotation and its orbit around the Earth result in a phenomenon called tidal locking, where the same side of the Moon always faces Earth. While one hemisphere is sometimes referred to as the "far side" or "dark side," it is important to note that it experiences both periods of sunlight and darkness just like the near side we see from Earth. The misconception arises because we cannot directly observe the far side from our vantage point on Earth, making it appear mysterious and unfamiliar to us.

Fact 97: Golf on the Moon.

During the Apollo missions, particularly Apollo 14 in 1971, astronaut Alan Shepard became the first and only person to play golf on the Moon. Shepard brought a modified golf club head and two golf balls with him on the mission. With the Moon's lower gravity, Shepard was able to hit the golf balls farther than he could on Earth. The playful act of hitting golf balls on the lunar surface added a unique and lighthearted touch to the historic Apollo missions, showcasing the ingenuity and human spirit of exploration.

Fact 98: The longest Consecutive stay in Space.

Valeri Polyakov holds the record for the longest consecutive stay in space. From January 8, 1994, to March 22, 1995, he spent a staggering 437 days living and working aboard the Mir space station. Polyakov's extended mission was part of a joint collaboration between Russia and NASA to study the effects of long-duration space travel on the human body. His remarkable endurance and resilience in the harsh environment of space provided valuable insights into the physiological and psychological challenges of prolonged space missions, paving the way for future endeavors beyond Earth's orbit.

Fact 99: The Sun takes 240 million years to complete an orbit of the Milky Way Galaxy.

Our Sun, along with the rest of the solar system, is not stationary but constantly in motion within the Milky Way Galaxy. It orbits around the galactic center, completing one revolution in approximately 240 million years. This vast cosmic timescale, known as a galactic year or cosmic year, highlights the immense size and slow-moving nature of galactic dynamics. As the Sun journeys through the galaxy, it encounters different regions, gravitational influences, and cosmic phenomena, contributing to the ever-changing cosmic tapestry of our galactic home.

Fact 100: Space travel has led to the invention and development of numerous technologies that we use in our daily lives.

The advancements made in space travel and exploration have had far-reaching impacts on technology and innovation. Many of the devices and technologies we rely on today have their origins in space-related research and development. For example, Space travel has led to inventions used in our daily lives, such as camera phones, LED lights, water filters, handheld vacuums, insulation, wireless headphones, and laptops. These technologies originated from space-related research and development, showcasing how space exploration has directly contributed to enhancing our everyday experiences.

CONTENTS:

CREDITS:

All images were provided by:

www.canva.com

NASA - National Aeronautics and Space
Administration

ESA - European Space Agency

Wikipedia

Author:
TOD.F

Printed in Great Britain
by Amazon